Node Localization in
Wireless Sensor Networks

Synthesis Lectures on Communications

Editor
William Tranter, *Virginia Tech*

Joint Source Channel Coding Using Arithmetic Codes
Dongsheng Bi, Michael W. Hoffman, and Khalid Sayood
2009

Fundamentals of Spread Spectrum Modulation
Rodger E. Ziemer
2007

Code Division Multiple Access(CDMA)
R. Michael Buehrer
2006

Game Theory for Wireless Engineers
Allen B. MacKenzie and Luiz A. DaSilva
2006

Node Localization in Wireless Sensor Networks

Xue Zhang, Cihan Tepedelenlioglu, Mahesh Banavar, and Andreas Spanias

ISBN: 978-3-031-00555-8   paperback
ISBN: 978-3-031-01683-7   ebook

DOI 10.1007/978-3-031-01683-7

A Publication in the Springer series
SYNTHESIS LECTURES ON COMMUNICATIONS

Lecture #12
Series Editor: William Tranter, *Virginia Tech*
Series ISSN
Print 1932-1244   Electronic 1932-1708

Node Localization in
Wireless Sensor Networks

Xue Zhang
Arizona State University

Cihan Tepedelenlioglu
Arizona State University

Mahesh Banavar
Clarkson University

Andreas Spanias
SenSIP Center, Arizona State University

SYNTHESIS LECTURES ON COMMUNICATIONS #12

ABSTRACT

In sensor network applications, measured data are often meaningful only when the location is accurately known. In this booklet, we study research problems associated with node localization in wireless sensor networks. We describe sensor network localization problems in terms of a detection and estimation framework and we emphasize specifically a cooperative process where sensors with known locations are used to localize nodes at unknown locations. In this class of problems, even if the location of a node is known, the wireless links and transmission modalities between two nodes may be unknown. In this case, sensor nodes are used to detect the location and estimate pertinent data transmission activities between nodes. In addition to the broader problem of sensor localization, this booklet studies also specific localization measurements such as time of arrival (TOA), received signal strength (RSS), and direction of arrival (DOA). The sequential localization algorithm, which uses a subset of sensor nodes to estimate nearby sensor nodes' locations is discussed in detail. Extensive bibliography is given for those readers who want to delve further into specific topics.

KEYWORDS

wireless sensor networks, location estimation, localization algorithms, GPS, DSP, Internet of Things (IoT)

Contents

Preface

The book was inspired by a series of projects on sensor networks in the SenSIP center and the School of Electrical, Computer and Energy Engineering at Arizona State University. The manuscript is intended to be an introduction to the topic of sensor localization. Much of the earlier work associated with this topic is collaborative among faculty and students. It started with the work in the group of Dr. Tepedelenlioglu followed later by work done in an NSF FRP grant to the SenSIP I/UCRC. This later gave rise to collaborative work between SenSIP researchers and Imperial College UDRC faculty funded in part by the British Council, which were followed by collaboration with Dr. Banavar at Clarkson University funded by a SenSIP I/UCRC grant on sensor localization. The specific focus of the book is based on Dr. Xue (Sophia) Zhang's dissertation which was completed under the supervision of Drs. Tepedelenlioglu and Spanias.

This book summarizes a series of sensor localization methods that can be used in wireless sensor networks and mobile communications. Chapter 1 introduces the topic and describes the use of anchors to localize sensors. Chapter 2 describes the performance of sensor location detection and estimation methods relative to the Cramer-Rao lower bound. Chapter 3 reviews linear and non linear sensor localization methods. Chapter 4 concentrates on sequential methods for localization. The book provides an extensive, though not exhaustive, list of references for the reader who wants to delve further into the details of each topic. Localization applications are also cited at the end of the book. Those include: internet of things, localization in indoor environments, acoustic sensing, underwater localization, smart homes, patient care, and emergency response.

Xue Zhang, Cihan Tepedelenlioglu, Mahesh Banavar, and Andreas Spanias
December 2016

Acknowledgments

The authors have been supported in part from the SenSIP center, the NCSS I/UCRC site and the NSF FRP award 123034.

CHAPTER 1

Introduction

1.1 WIRELESS SENSOR NETWORKS

Wireless sensor nodes (Figure 1.1) are typically compact and low power devices with an antenna, an embedded CPU, and a battery-powered radio. Sensor networks [1] consist of multiple sensor nodes that have a cluster of specialized transducers along with radio communications infrastructure intended to monitor and record data or events of interest. Sensors are often used to obtain measurements of location, temperature, humidity, irradiance, sound, and pressure [2–4]. A sensor network can be either wired or wireless although lately most applications typically involve radio communications where sensor nodes communicate with each other through a variety of wireless protocols (Fig. 1.2).

A wireless sensor network (WSN) can be either fully or partially connected; in the latter case, sensor nodes only communicate with a subset of nodes, usually neighboring nodes. A fully connected WSN benefits from network knowledge from the entire network where sensor nodes exchange information by transmitting and receiving signals from all other nodes, which can be

Figure 1.1: This sensor circuit was developed by Genetlab and has been used for intrusion detection in border and facility surveillance systems [1].

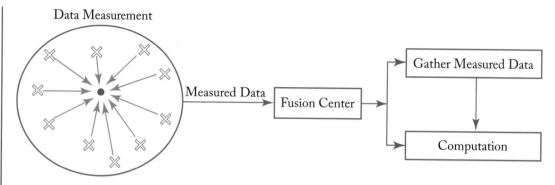

Figure 1.2: Wireless sensor network framework.

expensive in terms of computation, bandwidth, and power. On the other hand, in a partly connected WSN, each sensor node acquires information associated only with its nearest neighbors. Sensor networks can be either homogeneous, where all sensor nodes are identical in terms of battery life, communication range, and hardware complexity. On the other hand, in heterogeneous networks [5], sensor nodes have different communication ranges and functions.

Compared to traditional devices used in distributed sensing applications, the greatest advantages of WSNs are improved robustness and scalability [4]. Although the main driving forces for WSNs are fault tolerance, energy gain, and spatial capacity gain, WSNs have bandwidth [9] and power consumption limits [3]. More importantly, due to use on mobile platforms, one of the most important constraints on sensor nodes is low power consumption.

1.2   APPLICATIONS

WSNs have been deployed in both civilian and military applications. Applications of WSNs include: in security surveillance [10], health and wellness [6, 11], smart home [7, 12], Internet of Things (IoT) [71], fire protection in forests [13], and military tracking [8, 14]. Arampatzis et al. [15] provides a general literature review and extended bibliography addressing on the applications of WSNs, which include military applications, indoor monitoring, outdoor monitoring, and robotics. Applications in automobiles are described in Tavares et al. [16] while deployment of sensors in human health, medical care, and emergency rescue are found in Sun et al. [17]. In Yawut and Kilaso [18], the authors discuss WSN applications in weather and disaster alarm systems, and in Khedo et al. [19]. WSNs are applied in air pollution monitoring systems. Localization capabilities with cell phones have been of particular interest (see Figure 1.3), and mobile wireless sensor networks (MWSNs) in 4G and 5G systems have received a lot of attention recently, which are discussed in [141–145].

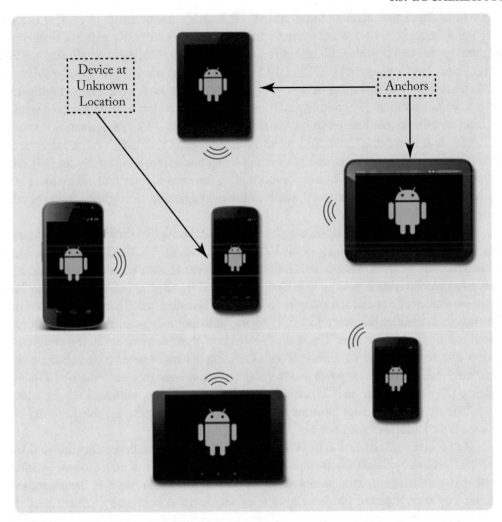

Figure 1.3: Mobile WSNs localization applications. By sending acoustic signals from anchors to the device with an unknown location, its location can be estimated [20].

1.3   LOCALIZATION

In 1973, the Global Positioning System (GPS), which is one of the most widely used technique for localization, was developed to overcome the limitations of previous navigation systems [21, 22]. GPS is been used in military, industry, and more recently, consumer/civilian applications. GPS provides location information in three dimensions and requires direct line-of-sight (LOS) with at least four satellites, providing an accuracy within 3 m. However, GPS does not work with obstacles that limit LOS communications between the satellites and the GPS receiver; therefore, its utility is limited in dense forests, mountains, and also in indoor environments [23–25]. Localization using radio frequencies overcomes some of these problems, particularly with the deployment of cellphone towers that provide sufficient information to localize mobile phones. In fact, the Federal Communications Commission (FCC) in the U.S. requires that wireless providers be able to locate mobile users within 10 m for 911 calls even in indoor environments with modest multipath [26]. A recent FCC document [27] highlights the need for increased localization accuracy when users call from mobile devices located in indoor environments. The FCC recommends that accurate localization is achieved in 30 s, within 3 m; however 90% of test calls have localization error greater than 100 m [28]. With a majority of 911 calls now being made from wireless devices [29], and over 56% of them being made from indoor locations [28], the need for accurate indoor localization is very important. A wrong estimate will result in the call appearing to originate from neighboring rooms, or even different floors.

To overcome GPS limitations, sensor networks can be applied for localization. Researchers have developed GPS-free techniques for locating nodes in WSNs. Sensors at known locations are designated as *anchors*, and are used to localize all other sensors which are at unknown locations, using methods such as time of arrival, or received signal strength. By measuring location related parameters, a node at an unknown location can be estimated. These methods can be scalable such as sequential discovery [30–32], where localization is performed using devices with limited communications range. The main drawback of these approaches is the requirement of accurate distance and direction information. Obtaining distance information in indoor environments involves specialized hardware and techniques such as web-access microwave [33], building modeling [34], antenna arrays [35], and ZigBee [36]. In addition, estimates require knowledge of parameters such as the exact transmit power and path-loss exponent for the medium [36], and clock synchronization between devices [20, 37].

The problem of localization in WSNs can be classified as location estimation and location detection. In some applications, anchors transmit signals to a node at an unknown location, and use their transmissions to localize the node. In other applications, such as fire protection in a building, one node is placed inside each room, whose location is known to all anchors. If a fire activates a node in any room, the active node at this known location needs to be detected. In the absence of transmission from a node, each anchor only receives noise, and each anchor receives a faded signal plus noise. Therefore, location detection is needed to decide whether a node is active or not. Localization in WSNs has been used in many applications, such as inventory tracking, forest fire tracking, home automation, and patient monitoring [38].

Localization problems where all the necessary computation takes place at a central node have a similar architecture to sensor networks with fusion centers [39, 40]. However, WSNs without a fusion center (fully distributed networks) have the advantage of robustness to node failures since they can function autonomously without a single node controlling the entire network [3]. In fully distributed networks, the sensors collaborate with their neighbors by exchanging information locally to come to an agreement on a global function of initial measurements. This agreement is often termed consensus [76]. Consensus is reached by each sensor for a sample statistic such as the sample mean.

The concept of cooperative WSNs relies on direct communication between nodes, which means nodes can communicate with each other. In localization problems, a node can estimate its location by sending or receiving signals from other nodes [41]. When anchors and other nodes communicate with the node that needs to be localized, a sensor network is called a cooperative WSN. In general, WSNs can be classified as cooperative and non-cooperative WSNs. On the other hand, in non-cooperative WSNs, no communications take place between nodes. Nodes can only communicate with anchors and estimate their locations through anchors. Figure 1.4 shows an example of cooperative WSNs. In the figure, nodes 1, 2, and 3 communicate with each other, which indicates that distance measurements d_{12}, d_{13}, and d_{23} are available. Figure 1.5 shows an example of non-cooperative WSNs. In the figure, the link between node 1 and node 2 is not present.

1.4 ORGANIZATION

The rest of the book is organized as follows. In Chapter 2, location estimation and detection are introduced, and the classification of localization algorithms are discussed. In Chapter 3, concepts associated with localization algorithms are reviewed, and the performance analysis on location estimation and detection is studied. Special emphasis is placed on the non-linear least squares, linear least squares, projection onto convex sets, and projection onto rings. In Chapter 4, a sequential location estimation scheme is discussed in detail, and compared with other localization algorithms.

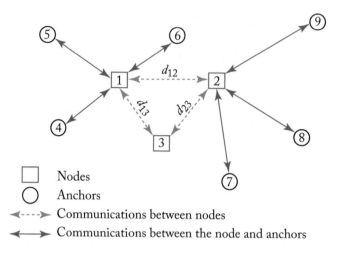

Figure 1.4: An example of cooperative WSNs. Here d_{12}, d_{13}, and d_{23} are the distances between nodes.

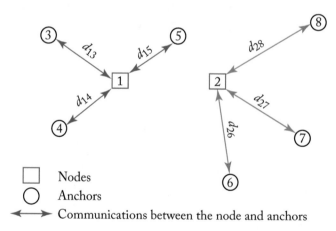

Figure 1.5: An example of non-cooperative WSNs. Here node 1 communicates with anchors 3, 4, and 5. Node 2 communicates with anchors 6, 7, and 8. Nodes do not communicate with each other.

CHAPTER 2

Introduction to Localization

To locate a node, a variety of localization algorithms that use linear and non linear techniques can be used. In this chapter, classification of localization algorithms and performance metrics, specifically, the Cramer-Rao lower bound (CRLB) are presented. In addition, probability of false alarm and probability of detection are introduced as metrics to evaluate the performance of detection algorithms.

2.1 CLASSIFICATION OF LOCALIZATION ALGORITHMS

Localization algorithms can be classified into three categories (see Figure 2.1). According to the computational capability at each anchor, localization algorithms can be classified as centralized and distributed algorithms. For centralized algorithms, shown in Figure 2.2, a fusion center (FC) is used to collect all information from sensor nodes. For distributed algorithms, as shown in Figure 2.3, each sensor node exchanges information with its neighbor or a group of nearby sensor nodes, and estimates parameters locally. An FC is optional, and if it exists, it is used to aggregate estimated parameters from each sensor node.

Centralized algorithms require more energy than distributed algorithms due to transmissions between sensor nodes and an FC. However, they provide more accurate results compared to distributed algorithms. In large WSNs, efficient utilization of energy is crucial for large area and long distance communications. In order to implement centralized algorithms more efficiently in large WSNs, many researchers focus on developing energy efficient protocol for WSNs [42, 43]. Comparing a distributed system with a centralized system, a distributed system is inherently more robust than a centralized system, due to lower possibility of link failures.

The sequential localization algorithm [30], attempts to overcome the limited power in WSNs by allowing nodes which have previously been localized to be used to localize other nodes [31]. Figure 2.4 shows an example of locating nodes using a sequential algorithm. In the figure, black nodes are at known or previously estimated locations. Once more nodes are determined, they become anchors to localize neighboring white nodes. The main drawback of this approach is that localization errors will propagate through the network during the iterative localization process. This is because it is assumed that the estimated locations of the nodes are the actual locations. However, due to the errors in localizing the nodes, this may not be the case. This makes the order in which nodes are localized important. In [31], the performance of different localization methods using sequential localization are compared.

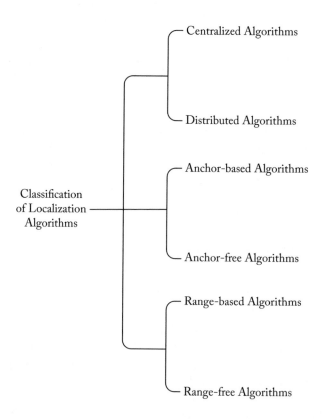

Figure 2.1: Classification of localization algorithms.

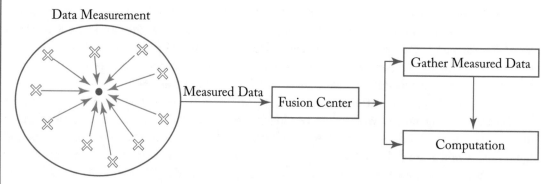

Figure 2.2: Flow chart for centralized algorithms. Sensor nodes collect measurements, then the measured data are passed to a FC. The fusion center is in charge of computing parameters of interest.

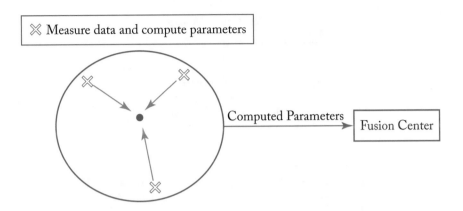

Figure 2.3: Flow chart for distributed algorithms. The measured data are used for computation at each sensor node.

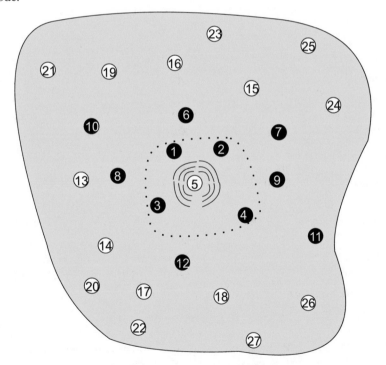

Figure 2.4: An example of locating nodes using the sequential algorithm. Black nodes are at known or previously estimated locations. Transmitting node 5 is localized using nodes 1, 2, 3, and 4. Once a white node is localized, it becomes an anchor to localize other neighboring white nodes [30].

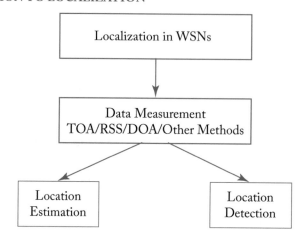

Figure 2.5: Flow chart of range-based localization.

Based on the need of anchors, algorithms can be classified as anchor-based and anchor-free algorithms. In the anchor-based scheme, several reference nodes, termed anchors, at known locations are used to localize nodes with unknown locations. In this scheme, the accuracy of the estimation depends on the number of anchors and performance is improved when more anchors are added to the network. In the anchor-free scheme, there is no anchor node with perfectly known location. Nodes communicate with each other to estimate relative locations instead of computing absolute locations [44]. Comparing the anchor-based scheme with the anchor-free scheme, the anchor-based scheme provides more accurate results than the anchor-free scheme. However, the hardware cost for the anchor-based scheme is much higher than the anchor-free scheme.

Localization schemes can also be classified as range-based and range-free approaches. In range-based localization schemes, location-related parameters are measured. Figure 2.5 shows the flow chart of the range-based localization approach. Commonly used range-based techniques include time of arrival (TOA) [45], received signal strength (RSS) [46], time difference of arrival (TDOA) [47], direction of arrival (DOA) [48], large aperture array (LAA) [30], and other hybrid techniques [49].

The TOA technique is one of the most popular techniques used for localization. Here, the time delay from the transmitting to the receiving node is measured, for either one-way or two-way transmission. For the one-way transmission, the time synchronization between the transmitter and the receiver is required [41, 47]. For two-way transmission schemes, there is no need for synchronization and the actual propagation time between the transmitter and receiver is half of the measured time. However, two-way transmission schemes require more energy and bandwidth compared to one-way methods.

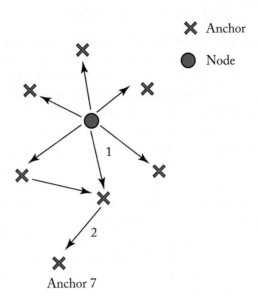

Figure 2.6: An example of range-free localization schemes.

In contrast to TOA, TDOA measures the difference between arrival times at the receiving nodes which removes the need for synchronization [50, 51]. However, this technique is known to suffer if there is insufficient bandwidth [37]. In RSS, a path loss model is used to estimate location on power loss measurements [52]. This is a simple and cheap technique to implement but suffers from problems in the presence of channel impairments such as multipath and frequency flat fading [53]. The DOA approach employs small aperture antenna arrays at each sensor node to estimate the direction of the transmitted signal. This method uses spatial diversity more optimally to achieve a better localization performance and does not require nodes to be synchronized but has increased processing requirements. Finally, in the LAA approach, nodes used for localization are aggregated to form large aperture array systems [30]. This approach is robust to frequency flat fading and may be extended to overcome co-channel interference.

Range-free algorithms include neighborhood and hop counting techniques. Commonly used range-free localization algorithms include DV-hop [54], in which each node counts the minimum number of hops to neighboring nodes, and estimates the distance by multiplying the number of hops with the average distance between two hops. Figure 2.6 shows an example of range-free localization algorithms. In this example, the number of hops between anchor 7 and the node is 2, and the number of hops between all other anchors and the node is 1. Range-based algorithms have higher accuracy compared to range-free methods, but require additional hardware.

2.2   ML ESTIMATOR FOR LOCATION

After location-related parameters are measured, the location can be computed. Several algorithms have previously been proposed to compute the node location [55–58, 72]; some provide the size of the entire network [73–76]. If the data is known to be described well by a particular statistical model, then the maximum likelihood estimator (MLE) can be derived and implemented [45]. Since the MLE can asymptotically achieve the CRLB, it is considered to be the optimal estimator. However, there are some difficulties with this approach. First, it is possible that optmization methods to find the MLE may determine local optima. Second, model-mismatch may occur, and results are no longer guaranteed to be optimal. Third, in a large WSN, finding a global optimum is computationally expensive. One way to prevent convergence to local optima is to formulate the location estimation as a convex optimization problem [55]. Convex constraints can be used such that a sensor's location estimate is required to be within a radius r from a second sensor. In [56], linear programming uses a "taxi metric" to provide a quick means to obtain rough localization estimates. More general constrains can be considered if semidefinite programming (SDP) techniques are used [57]. In [58] a distributed SDP-based localization algorithm was presented to simplify the complexity of computation.

2.3   PERFORMANCE ANALYSIS

To evaluate the performance of different estimation algorithms, the Cramer-Rao lower bound (CRLB), which provides a lower bound on the variance achievable by any unbiased location estimator, can be used to evaluate performance. An unbiased estimator will yield the true value of θ, on the average. Mathematically, an estimator is unbiased if

$$E(\hat{\theta}) = \theta. \tag{2.1}$$

The CRLB provides a lower bound on the variance achievable by the unbiased location estimator. Any unbiased estimator $\hat{\theta}$ must satisfy

$$\mathrm{cov}(\hat{\theta}) \geq \{\mathbb{E}[-\nabla_\theta(\nabla_\theta \log f(\mathbf{X}|\theta))^T]\}^{-1}, \tag{2.2}$$

where $f(\mathbf{X}|\theta)$ is the probability density function (pdf) of the observation $\mathbf{X}$, $\mathrm{var}(\hat{\theta})$ is the covariance of the estimator, and ∇_θ is the gradient operator w.r.t the vector θ. The Fisher information of the unknown parameter θ is defined as

$$\mathbf{F}_\theta = \mathbb{E}[-\nabla_\theta(\nabla_\theta \log f(\mathbf{X}|\theta))^T]. \tag{2.3}$$

The CRLBs on the TOA and RSS measurements in the absence of fading have been derived in Patwari et al. [41]. To locate a node in a 1 m by 1 m square, the CRLBs on the TOA measurement and the RSS measurement are shown in Figure 2.7 and Figure 2.8, respectively. From the figure one can see that the CRLB depends on the node location. When the node is located in the center of the square, the CRLB is the smallest in both cases.

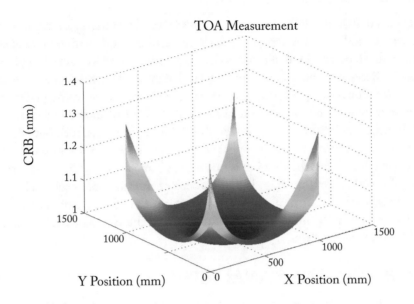

Figure 2.7: The CRLB on the TOA measurement when the node is located inside a 1 m by 1 m square.

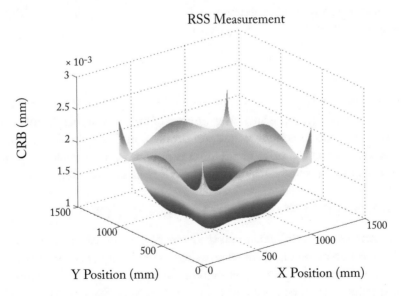

Figure 2.8: The CRLB on the RSS measurement when the node is located inside a 1 m by 1 m square.

2.4 LOCATION DETECTION

In some applications, such as in a surveillance network, the accuracy of the nodes's reading is crucial. Once any node behaves abnormal, the location of the node with abnormal activity needs to be estimated. To detect the malicious node, many researchers work on designing algorithms and protocols. Researches have been working on designing algorithms and protocols to detect abnormal nodes. Pries et al. [59] proposed detecting malicious nodes through detection of malicious message transmission in a network. In Ho et al. [60], a distributed node detection protocol is studied. In other applications, the node location is known to all anchors, but whether the node is active or not is unknown. In many applications such as detecting fire in buildings, each node is placed inside a room, and the location is known to all anchors. Anchors detect an event based on whether the node is transmitting. In the absence of transmission, each anchor receives only noise. Otherwise each anchor receives signal with noise. Therefore, location detection in WSNs is formulated as a binary hypothesis testing problem, and the Neyman-Pearson lemma [61] can be applied to solve the problem.

2.5 PERFORMANCE ANALYSIS

In detection theory, Neyman-Pearson hypothesis testing techniques are frequently used and the null hypothesis, states that the observed data only contains noise, and is denoted as H_0; another hypothesis states that the observed data contains both signal plus noise, and is denoted as H_1. The Neyman-Pearson detector, which is denoted as $L(\mathbf{x})$, is defined as the ratio of the log likelihood function under two hypotheses, which is given as

$$L(\mathbf{x}) = \frac{f(\mathbf{x}; H_1)}{f(\mathbf{x}; H_0)} \lessgtr \gamma. \tag{2.4}$$

It can be shown that the Neyman-Pearson detector is the optimal detector for maximizing the probability of detection ($\bar{P}_D$) while satisfying the constraint on the probability of false alarm ($\bar{P}_{FA}$). When the prior distribution is known, one can simply find the optimal threshold that satisfies constraints on $\bar{P}_{FA}$. When the prior distribution is unknown, machine learning algorithms can be applied to train the measured data and find the NP classifier [62, 63].

2.6 LOCATION ESTIMATION VS. DETECTION

The detection formulation is different from the estimation formulation in the following aspects. First of all, in detection problems, the goal is to detect the activity or silence of a node or multiple nodes at known locations; however, in estimation problems, the goal is to estimate the location of a node or multiple nodes, which are at unknown locations. Second, to estimate the location of a node, multiple anchors are needed in order to avoid ambiguity. For example, when using range-based methods, a minimum of two anchors are needed for one dimension (1D), and three anchors are needed for two dimensions (2D). On the other hand, to detect a node, each anchor can make a local decision on whether the node is active or not by correlating the received signal with the transmitted signal and then comparing with a threshold. The final decision can be made by exchanging this data with other anchors and an FC. Therefore, the detection problem can be solved by using a distributed implementation based on exchange of bits between the anchors and an FC. Third, performance analysis is different for these two formulations. In the estimation formulation, the variance of the location estimation error is used as a performance metric, whereas for detection, metrics such as the probability of false alarm and the probability of detection are used [61, 64].

CHAPTER 3

Review of Localization Algorithms

Once distance measurements are available, node location can be estimated using different approaches. In this chapter, existing approaches, such as nonlinear least squares (NLS), linear least squares (LLS), projection onto convex sets (POCS), and projection onto rings (POR) are described.

3.1 SYSTEM MODEL

Consider a 2D network with $N + M$ sensor nodes. Suppose that N nodes are placed at unknown locations $\mathbf{z}_i \in \mathbb{R}^2$, $i = 1, \ldots N$, and the remaining M nodes are anchors. Suppose that anchors are able to estimate distances to the nodes with the following observation:

$$\hat{d}_{ij} = d_{ij} + n_{ij}, \; j = 1, \ldots, M, i = 1, \ldots, N, \tag{3.1}$$

where $d_{ij} = \|\mathbf{z}_i - \mathbf{z}_j\|$ is the Euclidean distance between $\mathbf{z}_i$ and $\mathbf{z}_j$ and n_{ij} is the measurement error. We assume the measurement errors are independent and identically distributed (i.i.d). In what follows we describe several techniques to estimate the locations of the N nodes, based on the data observed according to (3.1).

3.2 NONLINEAR LEAST SQUARES

The nonlinear least squares location estimate based on the range measurement can be found as the solution to the non-convex optimization problem

$$\hat{\mathbf{z}}_i = \arg \min_{\mathbf{z}_i \in \mathbb{R}^2} \sum_{i=1}^{N} \sum_{j=1,\ldots,M} (\hat{d}_{ij} - d_{ij})^2, \tag{3.2}$$

and is described in Algorithm 3.1.

We note if n_{ij} are identically distributed, zero-mean Gaussian random variables for all $j = 1, \ldots, M$, the NLS estimate is also the maximum likelihood estimate [65]. Solving the NLS problems requires minimizing a nonlinear and non-convex function, which cannot be solved analytically. Therefore, numerical algorithms are applied to approximate NLS estimations. However, numerical algorithms require intensive computation and proper initialization, in order to obtain

---

Algorithm 3.1 NLS

---

1. Initialization: choose arbitrary initial target position $\mathbf{z}_i^0 \in \mathbb{R}^2$ for node i
2. for $k = 0$ until convergence or predefined number of K do
3. Update:

$$\hat{\mathbf{z}}_i = \arg \min_{\mathbf{z}_i \in \mathbb{R}^2} \sum_{j \in M} (\hat{d}_{i,j} - \|\mathbf{z}_i - \mathbf{z}_j\|)^2$$

4. End for

---

a closed form solution, location estimation problems can be solved using the linear least squares approach.

3.3   LINEAR LEAST SQUARES

Based on Gezici et al. [38] and Yi and Zhou [66], an alternative approach to the NLS estimation is the LLS approach. In a LLS technique, a new measurement set is obtained from the measurements as follows.

If we let $\hat{d}_{ij}$ represent the distance estimate obtained from the ith TOA measurement, and M represent the total number of anchors, then we have:

$$\hat{d}_{ij}^2 = (x_j - x_i)^2 + (y_j - y_i)^2, \text{ for } j = 1, \ldots, M, \tag{3.3}$$

where each distance measurement is assumed to have a circle of uncertain region. Then one of the equations in (3.3), say the rth one, is subtracted from all of the other equations. After some manipulation, the following linear relation can be obtained:

$$\mathbf{A}\mathbf{l} = \mathbf{P} \tag{3.4}$$

where $\mathbf{l} = [\mathbf{x}\ \mathbf{y}]^T$,

$$\mathbf{A} = 2 \begin{bmatrix} x_1 - x_r & y_1 - y_r \\ \vdots & \vdots \\ x_{r-1} - x_r & y_{r-1} - y_r \\ x_{r+1} - x_r & y_{r+1} - y_r \\ \vdots & \vdots \\ x_M - x_r & y_M - y_r \end{bmatrix} \tag{3.5}$$

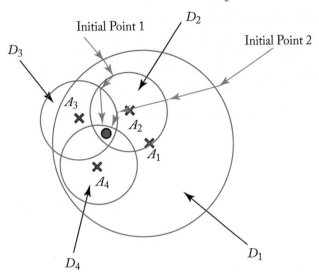

Figure 3.1: Projection onto convex sets.

and

$$\mathbf{P} = 2 \begin{bmatrix} d_r^2 - d_1^2 - k_r + k_1 \\ \vdots \\ d_r^2 - d_{r-1}^2 - k_r + k_{r-1} \\ d_r^2 - d_{r+1}^2 - k_r + k_{r+1} \\ \vdots \\ d_r^2 - d_M^2 - k_r + k_M \end{bmatrix} \tag{3.6}$$

with $k_j = x_j^2 + y_j^2$ and r being the selected reference node index that is used to obtain linear relations. Therefore, the LLS solution can be obtained using the pseudo-inverse as

$$\hat{\mathbf{I}} = (\mathbf{A}^T \mathbf{A})^{-1} \mathbf{A}^T \mathbf{P}. \tag{3.7}$$

This requires fewer computations relative to the NLS estimator. However, it is suboptimal in terms of the CRLB.

3.4   PROJECTION ONTO CONVEX SETS

To solve NLS-based problems using numerical approximations, proper initialization conditions are required. The projection onto Convex Sets (POCS) algorithm can provide good initialization and an accurate estimate of the node location. POCS was originally introduced to solve convex feasibility problems [67]. POCS has been applied to different problems in various fields, such as image restoration and radiation therapy treatment planning. Hero and Blatt [68] discussed the POCS in localization problems. In the absence of measurement error, it is clear that node i, at location $\mathbf{z}_i$, can be found in the intersection of a number of circles with radii d_{ij} and centers $\mathbf{z}_j$. For non-negative measurement errors, we can use discs because a target can be found inside the circles. We define the disc $\mathcal{D}_{ij}$ centered at $\mathbf{z}_j$ as

$$\mathcal{D}_{ij} = \{\mathbf{z}_i \in \mathbb{R}^2 : \|\mathbf{z}_i - \mathbf{z}_j\| \leq \hat{d}_{ij}\}, \ j = 1, \ldots, M. \tag{3.8}$$

Define an estimate of $\mathbf{z}_i$ as a point in the intersection $\mathcal{D}_i$ of the disc $\mathcal{D}_{ij}$

$$\hat{\mathbf{z}}_i \in \mathcal{D}_i = \bigcap_{j=1,\ldots,M} \mathcal{D}_{ij}. \tag{3.9}$$

Therefore, the positioning problem can be transformed to the following convex feasibility problem:

$$\text{find } \mathbf{z} = [\mathbf{z}_1, \ldots, \mathbf{z}_N] \text{ such that } \mathbf{z}_i \in \mathcal{D}_i, \ i = 1, \ldots, N. \tag{3.10}$$

Algorithm 3.2 POCS

---

1. Initialization: choose arbitrary initial target position $\mathbf{z}_i^0 \in \mathbb{R}^2$ for node i
2. for $k > 0$ until convergence or predefined number of K do
3. Update:

$$\mathbf{z}_i^{k+1} = \mathbf{z}_i^k + \lambda_k^i (\mathcal{P}_{\mathcal{D}_{ij}(k)}(\mathbf{z}_i^k) - \mathbf{z}_i^k)$$

4. End for

In Algorithm 3.2, we introduced $\mathcal{P}_{\mathcal{D}_{ij}(k)}$, which is the orthogonal projection of $\mathbf{z}$ onto set $\mathcal{D}_{ij}$. To find the projection of a point $\mathbf{z} \in \mathbb{R}^n$ onto a closed convex set $\Omega \subseteq \mathbb{R}^n$, we need to solve an optimization problem:

$$\mathcal{P}_\Omega(\mathbf{z}) = \arg\min \|\mathbf{z} - \mathbf{x}\|. \tag{3.11}$$

When Ω is a disc, there is a closed-form solution for the projection:

$$\mathcal{P}_{\mathcal{D}_{ij}}(\mathbf{z}) = \begin{cases} \mathbf{z}_j + \frac{\mathbf{z} - \mathbf{z_j}}{\|\mathbf{z} - \mathbf{z}_j\|} \hat{d}_{ij} & \|\mathbf{z} - \mathbf{z_j}\| \geq \hat{d}_{ij} \\ \mathbf{z} & \|\mathbf{z} - \mathbf{z}_j\| \leq \hat{d}_{ij}. \end{cases} \tag{3.12}$$

3.5   PROJECTION ONTO RINGS

In the case when the measurement noise is small, we can often improve POCS by replacing the disc $\mathcal{D}_{ij}$ with a ring defined as

$$\mathcal{R}_{ij} = \{\mathbf{x} \in \mathbb{R}^2 : \hat{d}_{ij} - \epsilon_l \leq d_j(\mathbf{z}) \leq \hat{d}_{ij} + \epsilon_u\}, \ j = 1, \ldots, M, \tag{3.13}$$

where $\epsilon_l + \epsilon_u$ determines the width of the ring. The width is a tuning parameter of the resulting algorithm. The projection onto rings (POR) is computed as in Algorithm 3.3.

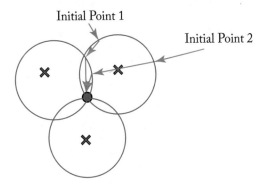

Figure 3.2: Projection onto rings.

Algorithm 3.3 POR

---

1. Initialization: choose arbitrary initial target position $\mathbf{z}_i^0 \in \mathbb{R}^2$ for node i
2. for $k > 0$ until convergence or predefined number of K do
3. Update:
$$\mathbf{z}_i^{k+1} = \mathbf{z}_i^k + \lambda_k^i (\mathcal{P}_{\mathcal{R}_{ij}(k)}(\mathbf{z}_i^k) - \mathbf{z}_i^k)$$

4. End for

CHAPTER 4

Sequential Localization

In large WSNs, centralized localization algorithms, which require anchors to transmit measurements to a fusion center, are hard to implement. On the other hand, distributed algorithms, such as the sequential localization algorithm, are more power efficient. However, localization errors will propagate through the network during the iterative localization process. This is because it is assumed that the estimated locations of nodes are the actual locations. In this chapter, the TOA, TDOA, RSS, DOA, and LAA localization algorithms will be expressed as the solution to a set of linear equations. Following this, the performance of a sequential network discovery process when using these different localization algorithms will be compared under different SNR regimes.

4.1 SYSTEM MODEL

Consider a homogeneous wireless sensor network with a large number of nodes at random unknown locations in $\mathbb{R}^2$ space. Assume that a small number of anchors at known locations are included in this network. Specifically, we assume the initial number of anchors is M, and the initial number of nodes is N, where $N >> M$. Each node or anchor has a circular coverage area with radius r, and operates at the frequency F_c. The Cartesian coordinate of the ith node is denoted by $\mathbf{z}_i$. The locations of all nodes at unknown locations using different localization algorithms. Each algorithm requires a different minimum number of anchors or previously localized nodes for localization to take place. In $\mathbb{R}^2$ space, for DOA, $M_{\min} = 2$, for TOA, TDOA, and RSS, $M_{\min} = 3$ and for LAA, $M_{\min} = 4$.

   The sequential location estimation process used in this chapter is described in Willerton et al. [30] and may be summarized as follows. Initially, one node transmits at an unknown location in the coverage area of the anchor nodes. If $M_{\min}$ nodes (anchors or the previously localized nodes) are within the coverage area of the node then the node location is estimated. If more than $M_{\min}$ nodes are available, all the available nodes will be used for localization. Otherwise, this node is skipped. Following this, another node transmits and the process continues with more nodes at estimated locations being available to perform localization of other nodes. Once localization has been attempted at all nodes, the process is repeated from the beginning, a total of K times. This gives an opportunity for skipped nodes to be revisited in so that more nodes in the vicinity of the troublesome nodes will now be at estimated locations to meet the $M_{\min}$ node requirement. In addition, nodes whose locations were successfully estimated may be gradually refined using data fusion techniques (including simple schemes such as averaging).

As errors in the sequential localization algorithm propagate during the estimation process, the choice of the first nodes becomes important. As the anchors will be more likely within the limited coverage area of the closest transmitting nodes, these closest nodes will be localized first. Multiple location estimates are needed to diminish the effect of localization order and noise.

4.2   LEAST SQUARES SOLUTIONS

We define $\mathbf{z}$ as the location of a node that needs to be localized. The location of the node may be estimated using measurements from a set of anchors, located at $\mathbf{p}_1, \mathbf{p}_2, \cdots, \mathbf{p}_M$, by solving the following set of linear equations:

$$\mathbb{H}\mathbf{z} = \mathbf{b}, \tag{4.1}$$

where the matrix $\mathbb{H}$ and the vector $\mathbf{b}$ depend upon the geometry of the receiving anchors with respect to the transmitting node, as well as metrics constructed from the data that nodes receive. Hence, (4.1) provides a unified notation for different localization approaches. Table 4.1 shows the structure of $\mathbb{H}$ and $\mathbf{b}$ for a selection of common localization algorithms in $\mathbb{R}^2$. For the TOA approach, $\hat{\tau}_i$ represents the estimated propagation time of the line-of-sight (LOS) signal between the node (at $\mathbf{z}$) and the ith anchor. For the TDOA approach, $\hat{\tau}_{i1}$ is the time difference from LOS signal at the 1st anchor to the ith anchor. For the RSS approach, $\hat{d}_i^2$ is the estimated distance between the ith anchor and the node. For the DOA approach, θ_i and ρ_i represent the angle and range associated with the ith array which has an array reference point at $\mathbf{p}_i$. The angles are estimated using a DOA algorithm (e.g., MUSIC [69]), and the ranges to the source may be easily inferred via the sine rule using these estimated directions of arrival and the locations of the array reference points. For example, for the triangle defined by the points $\mathbf{p}_1$, $\mathbf{p}_2$, and $\mathbf{p}_3$,

$$\rho_1 = \frac{\|\mathbf{p}_2 - \mathbf{p}_1\| \sin(\pi - \theta_2 + \psi_{12})}{\sin(\theta_2 - \theta_1)}, \tag{4.2}$$

$$\rho_2 = \frac{\|\mathbf{p}_2 - \mathbf{p}_1\| \sin(\theta_1 - \psi_{12})}{\sin(\theta_2 - \theta_1)}, \tag{4.3}$$

where ψ_{12} is the known angle between the first and second array reference points. Finally, for the LAA approach, $\mathbf{z}$ from (4.1) is a four element evector with the final element equal to $\|\mathbf{z}\|^2$ and $\underline{\mathcal{K}} \in \mathbb{R}^{(M-1) \times 1}$ is constructed by rotating the array reference point [30, 45, 70] with reference to a global reference point set at the origin. Note that this is the only approach which uses sensors collectively as a single array system [30, 92].

4.3   PERFORMANCE COMPARISON OF SEQUENTIAL LOCATION ESTIMATION

Consider a two-dimensional 600 m ×600 m sensor field, as shown in Figure 4.5. Assume $M = 4$ anchors denoted by blue triangles are placed at known locations in the center of the field. These

Table 4.1: Linear equations for node localization using TOA, TDOA, RSS, DOA, and LAA techniques in $\mathbb{R}^2$ space (see Equation (4.1) and Figures 4.1–4.4)

<u>TOA</u>

$$\mathbb{H} = [\mathbf{p}_2 - \mathbf{p}_1, \mathbf{p}_3 - \mathbf{p}_1, \cdots, \mathbf{p}_M - \mathbf{p}_1]^T$$
$$\in \mathbb{R}^{(M-1)\times 2}$$

$$\mathbf{b} = \frac{1}{2} \begin{bmatrix} \|\mathbf{p}_2\|^2 - \|\mathbf{p}_1\|^2 - c^2\left(\hat{\tau}_2^2 - \hat{\tau}_1^2\right) \\ \|\mathbf{p}_3\|^2 - \|\mathbf{p}_1\|^2 - c^2\left(\hat{\tau}_3^2 - \hat{\tau}_{10}^2\right) \\ \vdots \\ \|\mathbf{p}_M\|^2 - \|\mathbf{p}_1\|^2 - c^2\left(\hat{\tau}_M^2 - \hat{\tau}_1^2\right) \end{bmatrix}$$
$$\in \mathbb{R}^{(M-1)\times 1}$$

<u>TDOA</u>

$$\mathbb{H} = [\mathbf{p}_2 - \mathbf{p}_1, \mathbf{p}_3 - \mathbf{p}_1, \cdots, \mathbf{p}_M - \mathbf{p}_1]^T$$
$$\in \mathbb{R}^{(M-1)\times 2}$$

$$\mathbf{b} = \frac{1}{2} \begin{bmatrix} \|\mathbf{p}_2\|^2 - \|\mathbf{p}_1\|^2 - c^2\left(\hat{\tau}_{21}^2 + 2\hat{\tau}_1 \cdot \hat{\tau}_{21}\right) \\ \|\mathbf{p}_3\|^2 - \|\mathbf{p}_1\|^2 - c^2\left(\hat{\tau}_{31}^2 + 2\hat{\tau}_1 \cdot \hat{\tau}_{31}\right) \\ \vdots \\ \|\mathbf{p}_M\|^2 - \|\mathbf{p}_1\|^2 - c^2\left(\hat{\tau}_{M1}^2 + 2\hat{\tau}_1 \cdot \hat{\tau}_{M1}\right) \end{bmatrix}$$
$$\in \mathbb{R}^{(M-1)\times 1}$$

<u>RSS</u>

$$\mathbb{H} = [\mathbf{p}_2 - \mathbf{p}_1, \mathbf{p}_3 - \mathbf{p}_1, \cdots, \mathbf{p}_M - \mathbf{p}_1]^T$$
$$\in \mathbb{R}^{(M-1)\times 2}$$

$$\mathbf{b} = \frac{1}{2} \begin{bmatrix} \|\mathbf{p}_2\|^2 - \|\mathbf{p}_1\|^2 - \left(\hat{d}_2^2 - \hat{d}_1^2\right) \\ \|\mathbf{p}_3\|^2 - \|\mathbf{p}_1\|^2 - \left(\hat{d}_3^2 - \hat{d}_1^2\right) \\ \vdots \\ \|\mathbf{p}_M\|^2 - \|\mathbf{p}_1\|^2 - \left(\hat{d}_M^2 - \hat{d}_1^2\right) \end{bmatrix}$$
$$\in \mathbb{R}^{(M-1)\times 1}$$

<u>DOA</u>

$$\mathbb{H} = \mathbf{p}_1 \otimes \mathbb{I}_2$$
$$\in \mathbb{R}^{2M\times 2}$$

$$\mathbf{b} = \begin{bmatrix} \mathbf{p}_1 + \rho_1 \cdot [\cos\theta_1, \sin\theta_1]^T \\ \mathbf{p}_2 + \rho_2 \cdot [\cos\theta_2, \sin\theta_2]^T \\ \vdots \\ \mathbf{p}_M + \rho_M \cdot [\cos\theta_M, \sin\theta_M]^T \end{bmatrix}$$
$$\in \mathbb{R}^{2M\times 1}$$

<u>LAA</u>

$$\mathbb{H} = \begin{bmatrix} 2\left(\mathbb{1}_{M-1}\mathbf{p}_1^T - [\mathbf{p}_2, \mathbf{p}_3, \cdots, \mathbf{p}_M]^T\right) \\ \left(\mathbb{1}_{M-1} - \underline{\mathcal{K}}^2\right) \end{bmatrix}$$
$$\in \mathbb{R}^{(M-1)\times 2}$$

$$\mathbf{b} = \left[\|\mathbf{p}_1\|^2 \mathbb{1}_{M-1} - \left[\|\mathbf{p}_2\|^2, \|\mathbf{p}_3\|^2, \cdots, \|\mathbf{p}_M\|^2\right]^T\right]$$
$$\in \mathbb{R}^{(M-1)\times 1}$$

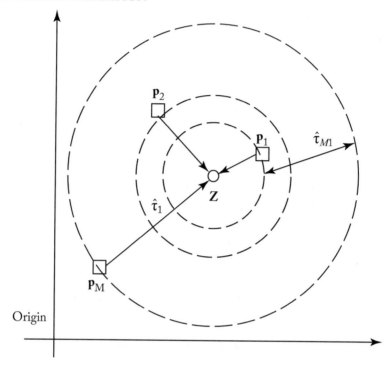

Figure 4.1: TOA/TDOA-based localization of a node located at $\mathbf{z}$ using M anchors at locations $\mathbf{p}_1, \mathbf{p}_2, \ldots, \mathbf{p}_M$.

anchors are surrounded by 200 sensor nodes denoted by green circles placed randomly at unknown locations according to a 2D uniform distribution. All nodes transmit at a frequency of $F_c = 2.45\,\text{GHz}$. Each node has a transmission range of $r = 100\,\text{m}$ and transmits $L = 100$ snapshots of sinusoidal signals. A propagation constant of $a = 4$ exists in the simulation environment and it is assumed that their is no flat frequency fading or multipath. Localization of each transmitting node is attempted $K = 200$ times, and if no location estimate was determined it is declared undiscovered. In Figure 4.5, the results of the sequential discovery process using LAA localization are shown. Estimated locations are represented using red crosses. Two of the nodes remain undiscovered because there are not enough nodes in their coverage range for localization.

Figure 4.6 shows the performance of TOA, TDOA, RSS, DOA, and LAA techniques. RMSE values are plotted for different values of $\text{SNR} \times L$. From the figure one can see that for all techniques, the average RMSE decreases as $\text{SNR} \times L$ increases. This is expected as noise reduces and more data is available to produce more statistically efficient metrics. When $\text{SNR} \times L$ is below 70 dB, DOA has the best performance. However, when the $\text{SNR} \times L$ is greater than 70 dB, LAA exceeds its performance. Note that the TOA and TDOA schemes are markedly poor due

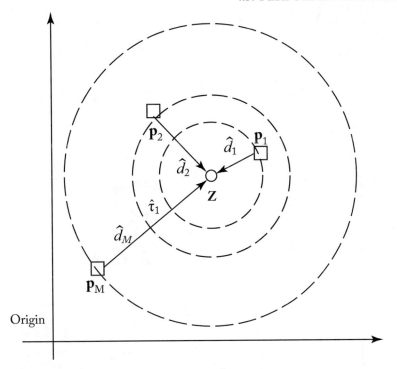

Figure 4.2: RSS-based localization of a node at $\mathbf{z}$ using M anchors at $\mathbf{p}_1, \mathbf{p}_2, \ldots, \mathbf{p}_M$.

to bandwidth limitations. While the performance of the RSS regime is good in this simulation environment, with more complex channel effects, performance will rapidly degrade.

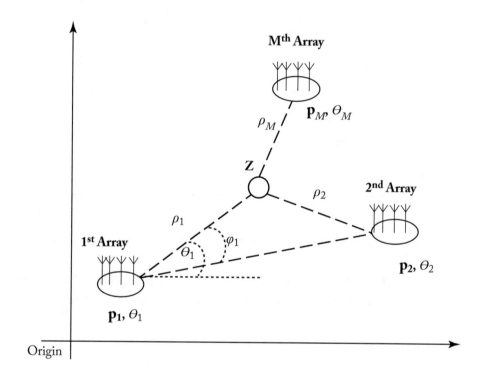

Figure 4.3: DOA-based localization of a node located at $\mathbf{z}$ using M arrays with local geometries $\mathbf{p}_1, \mathbf{p}_2, \ldots, \mathbf{p}_M$ and DOAs $\theta_1, \theta_2, \ldots, \theta_M$.

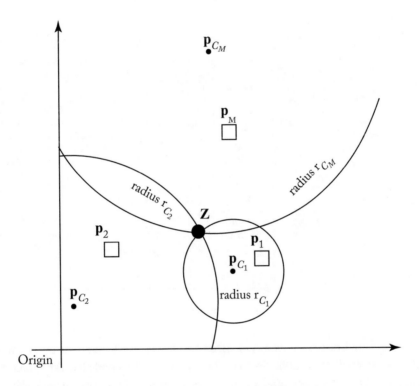

Figure 4.4: LAA-based localization of a node located at **z** using M anchors at known locations $\mathbf{p}_1, \mathbf{p}_2, \ldots, \mathbf{p}_M$ and AOAs $\theta_1, \theta_2, \ldots, \theta_M$. Here, r_{c_i} and $\mathbf{p}_{c_j}$ denote the radius and center of the jth circular locus which may be used to estimate the node location.

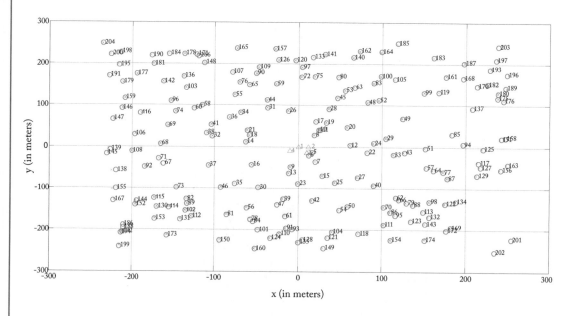

Figure 4.5: Sequential discovery using the LAA technique. Each node having a coverage of $r = 100$ m and SNR $= 30$ dB; $K = 200$. Green circles represent actual node locations, blue triangles represent anchor nodes at known locations, and red crosses represent location estimates. All but nodes 201 and 202 are localized. Location uncertainties can be seen more predominantly at the edges of the network.

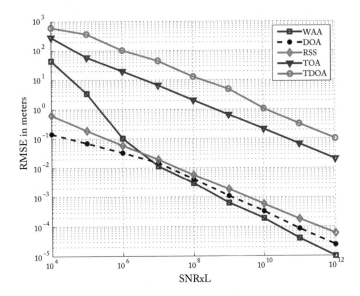

Figure 4.6: Performance comparison of different localization techniques for network discovery. System parameters are $r = 100$ m; $L = 100$; $K = 200$. Average RMSE is plotted vs. SNR×L.

CHAPTER 5

Concluding Remarks and Summary

The book focused on concepts surrounding wireless sensor networks and more specifically node localization approaches. In Chapter 1, we introduced the basic concepts, notation, and applications. We also discussed various topologies for sensor node localization. We then described the problem of sensor localization, introduced the idea of establishing and using anchors, and we briefly surveyed methods for estimating the location of the nodes. In our introductory remarks we distinguished the problems of node detection and node location estimation, and we also introduced the concept of centralized estimation of parameters using sensor fusion methods. The notion of sensor node counting [74, 128] and sensor network size was also introduced along with concepts of consensus based estimation [75, 76, 126, 127]. Sequential discovery was proposed for nodes that have limited transmission power and range.

In Chapter 2, we concentrated on location detection and estimation methods and described the basics of performance analysis in sensor localization applications. We introduced the Cramer-Rao lower bound which we used to analyze and compare different localization approaches. We distinguished several methods including centralized, distributed, range-based, and anchor-based localization algorithms. We also discussed briefly direction of arrival, time of arrival methods, as well as received signal strength (RSS). We reviewed maximum likelihood estimators and their ability to asymptotically achieve the CRLB. We showed sample results for TOA and RSS measurements relative to the CRLB. Finally, we discussed the difference in the formulation of location detection and estimation.

In Chapter 3, we reviewed several localization algorithms with the emphasis on linear and nonlinear methods. We first introduced the model and topology for location estimation and we discussed methods that use anchors to determine the sensor location. Nonlinear least squares formulations were given based on range measurements and an algorithm is outlined. Linear least squares methods were also explained in Chapter 3 using diagrams and equations, and the solution to the LLS using a pseudo-inverse is described. Solutions using projections on convex sets were described and iterative methods (POCS) and equations were given. In addition, we provided a section on projections on rings (POR) and again an iterative algorithm was given.

In Chapter 4, we discussed sequential methods for localization and we provided arguments on how error in sequential estimation propagates through the network. We explained the importance of the initial estimates and the significance of defining accurately the anchors. Least squares solutions were reviewed and TOA and DOA approaches were described. Linear equations for several methods including TOA, TDOA, DOA, RSS and LAA were summarized in Table 4.1. Performance comparisons were given for all of these methods for different signal to noise ratios. Sequential discovery was simulated and results were obtained using the LAA technique. Performance comparisons were also given in terms of the RMSE and the SNR at the end of the chapter.

The field of sensor node localization is vast and the reader is encouraged to delve further in bibliography to study in depth the various methods in terms of performance and complexity. For this reason we provide additional references in this book to assist the reader in examining problems and solutions for various sensor topologies and noise conditions. Some of the well-cited papers and surveys are given in [3, 77–81, 117–120, 152–156]. DOA methods are described in several papers for a variety of sensor and antenna applications [82–92, 148]. Additional references on TOA and TDOA are given in [93–98]. Kalman based methods are given in [129–131]. Distributed localization [99, 123–125] and RSS localization related articles are given in [100–102]. Application articles for sensor localization include: internet of things [103–105], localization in indoor environments [106, 107, 149–151], acoustic sensing [108, 109, 132–137], underwater localization [110], robots [138, 139], military [146, 147], smart homes [111], RFID [121, 122], rescue assistance [112], Alzheimer patient care [113], and emergency response [114–116].

Bibliography

[1] Wireless Sensor Node, Genetlab Bilgi Teknolojileri San. Ve Tic. A.S. 1

[2] D. Culler, D. Estrin, and M. Sivastava, Overview of sensor networks, *IEEE Computer Society*, pp. 41–49, August 2004. 1

[3] I. Akyildiz, W. Su, Y. Sankarasubramaniam, and E. Cayirci, A survey on sensor networks, *IEEE Communications Magazine*, vol. 40, no. 8, pp. 102–114, August 2002. DOI: 10.1109/mcom.2002.1024422. 2, 4, 34

[4] J. Yick, B. Mukherjee, and D. Ghosal, Wireless sensor network survey, *Computer Networks*, vol. 52, no. 12, pp. 2292–2330, August 2008. DOI: 10.1016/j.comnet.2008.04.002. 1, 2

[5] V. Mhatre, Homogeneous vs. heterogeneous clustered sensor networks: A comparative study, in *IEEE International Conference on Communications*, pp. 3646–3651, June 2004. DOI: 10.1109/ICC.1995.524186. 2

[6] R. Frank, Sensor advance medical and healthcare applications. [Online]. http://www.designworldonline.com/sensors-advance-medical-and-healthcare-applications 2

[7] Logical neighborhoods: programming wireless sensor networks. [Online]. http://logicalneighbor.sourceforge.net 2

[8] Introduction to wireless sensor networks and its applications. [Online]. https://wirelessmeshsensornetworks.wordpress.com/tag/wireless-sensor-network-technology-and-its-application-using-vlsi 2

[9] U. Bilstrup, K. Sjoberg, B. Svensson, and P. Wiberg, Capacity limitations in wireless sensor networks, *IEEE Emerging Technologies and Factory Automation*, pp. 529–536, September 2003. DOI: 10.1109/etfa.2003.1247752. 2

[10] A. Chehri, P. Fortier, and P. Tardif, Security monitoring using wireless sensor networks, *IEEE Annual Conference on Communication Networks and Services Research*, pp. 13–17, May 2007. DOI: 10.1109/cnsr.2007.58. 2

[11] J. Ko, C. Lu, M. Srivastava, J. Stankovic, A. Terzis, and M. Welsh, Wireless sensor networks for healthcare, *IEEE Proceedings*, pp. 1947–1960, September 2010. DOI: 10.1109/jproc.2010.2065210. 2

[12] D. Basu, G. Moretti, G. Gupta, and S. Marsland, Wireless sensor network based smart home: sensor selection, deployment and monitoring, *IEEE Sensors Applications Symposium*, pp. 49–54, February 2013. DOI: 10.1109/sas.2013.6493555. 2

[13] J. Zhang, W. Li, Z. Yin, S. Liu, and X. Guo, Forest fire detection system based on wireless sensor network, *IEEE Conference on Industrial Electronics and Applications*, pp. 520–523, May 2009. DOI: 10.1109/iciea.2009.5138260. 2

[14] T. Alhmiedat, A. Taleb, and M. Bsoul, A study on threats detection and tracking systems for military applications using WSNs, *International Journal of Computer Applications*, vol. 40, no. 15, pp. 12–18, February 2012. DOI: 10.5120/5055-7347. 2

[15] T. Arampatzis, J. Lygeros, and S. Manesis, A survey of applications of wireless sensors and wireless sensor networks, *13th Mediterranean Conference on Control and Automation*, pp. 719–724, June 2005. DOI: 10.1109/.2005.1467103. 2

[16] J. Tavares, F. Velez, and J. Ferro, Application of wireless sensor networks to automobiles, *Measurement Science Review*, vol. 8, no. 3, pp. 65–71, 2008. DOI: 10.2478/v10048-008-0017-8. 2

[17] G. Sun, J. Chen, W. Guo, and K. Liu, Signal processing techniques in network-aided positioning: A survey of state-of-the-art positioning designs, *IEEE Signal Processing Magazine*, vol. 22, no. 4, pp. 12–23, June 2005. DOI: 10.1109/msp.2005.1458273. 2

[18] C. Yawut and S. Kilaso, A wireless sensor network for weather and disaster alarm systems, *2011 International Conference on Information and Electronics Engineering*, vol. 6, pp. 155–159, 2011. 2

[19] K. Khedo, R. Perseedoss, and A. Mungur, A wireless sensor network air pollution monitoring system, *International Journal of Wireless and Mobile Networks*, vol. 2, no. 2, pp. 31–45, 2010. DOI: 10.5121/ijwmn.2010.2203. 2

[20] P. Curtis, M. Banavar, S. Zhang, A. Spanias, and V. Weber, Android acoustic ranging, in *IEEE International Conference on Information, Intelligence, Systems, and Applications*, pp. 118–123, Crete, July 2014. DOI: 10.1109/iisa.2014.6878721. 3, 4

[21] E. Kaplan, *Understanding GPS: Principles and Applications*. Artech House Telecommunication Library, 1996. 3

[22] S. Miller, X. Zhang, and A. Spanias, A new asymmetric correlation kernel for gnss multipath mitigation, *IEEE Sensor Signal Processing for Defence*, pp. 1–5, Edinburgh, September 2015. DOI: 10.1109/sspd.2015.7288498. 3

[23] A. Pal, Localization algorithms in wireless sensor networks: Current approaches and future challenges, *Network Protocols and Algorithm*, vol. 2, no. 1, pp. 45–74, January 2010. DOI: 10.5296/npa.v2i1.279. 4

[24] Y. Shen and M. Win, Fundamental limits of wideband localization—Part I: A general framework, *IEEE Transactions on Information Theory*, pp. 4956–4980, September 2010. DOI: 10.1109/tit.2010.2060110.

[25] Y. Shen, H. Wymeersch, and M. Win, Fundamental limits of wideband localization—Part II: Coopeartive networks, *IEEE Transactions on Information Theory 2010*, pp. 4981–5000, June 2010. DOI: 10.1109/tit.2010.2059720. 4

[26] F. Commission, Revision of the commissions rules to insure compatibility with enhanced 911 emergency calling systems, *FCC Docket*, no. 94–102, December 1996. 4

[27] Wireless E911 location accuracy requirements, *Federal Communications Commission*, February 2014. 4

[28] E911 location accuracy: Indoor localization test bed report. [Online]. `http://transition.fcc.gov/bureaus/pshs/advisory/csric3/CSRIC_III_WG3_Re port_March_%202013_ILTestBedReport.pdf` 4

[29] For 911, is a cell phone as safe as a landline? [Online]. `http://www.consumerreports.or g/cro/magazine-archive/2011/january/electronics/best-cell-phones/911-from-cell-phone/index.htm` 4

[30] M. Willerton, M. Banavar, X. Zhang, A. Manikas, C. Tepedelenlioglu, A. Spanias, T. Thomoton, E. Yeatman, and A. Constantinides, Sequential wireless sensor network discovery using wide aperture array signal processing, *European Signal Processing Conference*, pp. 2278–2282, August 2012. 4, 7, 9, 10, 11, 23, 24

[31] X. Zhang, M. Banavar, M. Willerton, A. Manikas, C. Tepedelenlioğlu, A. Spanias, T. Thornton, E. Yeatman, and A. Constantinides, Performance comparison of localization thechniques for sequential WSN discovery, in *IEEE Sensor Signal Processing for Defence*, pp. 1–5, London, September 2012. DOI: 10.1049/ic.2012.0120. 7

[32] X. Zhang, C. Tepedelenlioglu, M. Banavar, and A. Spanias, Distributed location detection in wireless sensor networks, pp. 428–432, November 2013. DOI: 10.1109/acssc.2013.6810312. 4

[33] M. Yamamoto, T. Ohtsuki, H. Utsumi, and N. Furukawa, Mobile-phone indoor localization based on microwave identification using web-access time, *IEEE Topical Conference on Wireless Sensors and Sensor Networks (WiSNet)*, pp. 28–30, January 2014. DOI: 10.1109/wisnet.2014.6825494. 4

[34] N. Li, B. Becerik-Gerber, B. Krishnamachari, and L. Soibelman, A bim centered indoor localization algorithm to support building fire emergency response operations, *Automation in Construction*, vol. 42, pp. 78–89, June 2014. DOI: 10.1016/j.autcon.2014.02.019. 4

[35] X. Wang, S. Mao, S. Pandey, and P. Agrawal, CA2T: Cooperative antenna arrays technique for pinpoint indoor localization, *Proc. Computer Science*, vol. 34, pp. 392–399, 2014. DOI: 10.1016/j.procs.2014.07.044. 4

[36] F. Salim, M. Williams, N. Sony, M. Dela Pena, Y. Petrov, A. Saad, and B. Wu, Visualization of wireless sensor networks using zigbee's received signal strength indicator (RSSI) for indoor localization and tracking, *2014 IEEE International Conference on Pervasive Computing and Communications Workshops (PERCOM)*, pp. 575–580, March 2014. DOI: 10.1109/percomw.2014.6815270. 4

[37] P. Curtis, M. Banavar, V. Weber, and A. Spanias, Signals and systems demonstrations for undergraduates using android-based localization, *IEEE FIE*, 2014. DOI: 10.1109/fie.2014.7044095. 4, 11

[38] S. Gezici, I. Guvenc, and Z. Sahinoglu, On the performance of linear least-squares estimation in wireless positiong systems, *IEEE Communication Society*, pp. 4203–4208, May 2008. DOI: 10.1109/icc.2008.789. 4, 18

[39] D. Castanon and D. Teneketzis, Distributed estimation algorithms for nonlinear systems, *IEEE Transactions on Automatic Control*, vol. 30, no. 4, pp. 418–425, 1985. DOI: 10.1109/tac.1985.1103972. 4

[40] A. Alouani, Distributed estimation algorithms for nonlinear systems, *IEEE Transactions on Automatic Control*, vol. 35, no. 9, pp. 1078–1081, 1990. DOI: 10.1109/9.58543. 4

[41] N. Patwari, A. Hero, M. Perkins, N. Correal, and R. O'Dea, Relative location estimation in wireless sensor networks, *IEEE Transactions on Signal Processing*, vol. 51, no. 8, pp. 2137–2148, July 2003. DOI: 10.1109/tsp.2003.814469. 5, 10, 12

[42] S. Muruganathan, D. Ma, R. Bhasin, and A. Fapojuwo, A centralized energy-efficient routing protocol for wireless sensor networks, *IEEE Communications Society*, pp. 8–13, March 2005. DOI: 10.1109/mcom.2005.1404592. 7

[43] L. Lu and C. Lim, Position-based, energy-efficient, centralized clustering protocol for wireless sensor networks, *4th IEEE Conference on Industrial Electronics and Applications*, pp. 139–144, May 2009. DOI: 10.1109/iciea.2009.5138185. 7

[44] S. Shioda and K. Shimamura, Anchor-free localization: estimation of relative locations of sensors, *IEEE International Symposium on Personal, Indoor and Mobile Raido Communications: Mobile and Wireless Networks*, pp. 2087–2092, September 2013. DOI: 10.1109/pimrc.2013.6666488. 10

[45] N. Patwari, J. Ash, S. Kyperountas, A. Hero, R. Moses, and N. Correal, Locating the nodes: Cooperative localization in wireless sensor networks, *IEEE Signal Processing Magazine*, vol. 22, no. 4, pp. 54–59, June 2005. DOI: 10.1109/msp.2005.1458287. 10, 12, 24

[46] Y. Shen, Fundamental limits of wideband localization, Master's thesis, Massachusetts Institute of Technology, Feburary 2008. 10

[47] A. Catovic and Z. Sahinoglu, The Cramér-Rao bounds of hybrid TOA/RSS and TDOA/RSS location estimation schemes, *IEEE Communications Letters*, vol. 8, no. 10, pp. 626–628, October 2004. DOI: 10.1109/lcomm.2004.835319. 10

[48] J. Xu, M. Ma, and C. Law, AOA cooperative position localization, in *Global Telecommunications Conference, IEEE*, pp. 1–5, December 2008. DOI: 10.1109/glocom.2008.ecp.720. 10

[49] M. Laaraiedh, L. Yu, S. Avrillon, and B. Uguen, Comparison of hybrid localization schemes using RSSI, TOA, and TDOA, in *11th European Wireless Conference*, pp. 1–5, April 2011. 10

[50] S. Gezici, A survey on wireless position estimation, *Wireless Personal Communications*, vol. 44, no. 3, pp. 263–282, February 2008. DOI: 10.1007/s11277-007-9375-z. 11

[51] E. Huang and R. W. Herring, Comparisons of error characteristics between TOA and TDOA positioning, *IEEE Transactions on Acoustics, Speech, Signal Processing*, no. 1, pp. 307–311, October 1981. 11

[52] H. Wymeersch, J. Lien, and M. Z. Win, Cooperative localization in wireless networks, *Proc. of the IEEE*, vol. 97, no. 2, pp. 427–450, February 2009. DOI: 10.1109/jproc.2008.2008853. 11

[53] M. Win, A. Conti, S. Mazuelas, Y. Shen, W. Gifford, D. Dardari, and M. Chiani, Network localization and navigation via cooperation, *Communications Magazine, IEEE*, pp. 56–62, May 2011. DOI: 10.1109/mcom.2011.5762798. 11

[54] D. Niculescu and B. Nath, Ad-hoc positioning system, in *USENIX Technical Annual Conference*, pp. 317–327, June 2002. DOI: 10.1109/glocom.2001.965964. 11

[55] L. Doherty, K. Pister, and L. Ghaoui, Convex position estimation in wireless sensor networks, *IEEE Infocom*, pp. 1655–1663, April 2001. DOI: 10.1109/infcom.2001.916662. 12

[56] E. Larsson, Cramer-Rao bound analysis of distributed positioning in sensor networks, *IEEE Signal Processing Letters*, no. 3, pp. 334–337, March 2004. DOI: 10.1109/lsp.2003.822899. 12

[57] P. Biswas and Y. Ye, Semidefinite programming for ad hoc wireless sensor network localization, *IEEE Information Processing in Sensor Networks*, pp. 45–54, April 2004. DOI: 10.1145/984622.984630. 12

[58] P. Biswas and Y. Ye, A distributed method for solving semidefinite programming for ad hoc wireless sensor network localization, Stanford University, Tech. Rep., October 2003. DOI: 10.1007/0-387-29550-x_2. 12

[59] W. Pries, T. de Paula Figueiredo, H. Wong, and A. Loureiro, Malicious node detection in wireless sensor networks, *IEEE Parallel and Distributed Processing Symposium*, April 2004. DOI: 10.1109/ipdps.2004.1302934. 14

[60] J. Ho, M. Wright, and S. Das, Distributed detection of mobile malicious node attacks in wireless sensor networks, *Ad Hoc Networks*, pp. 512–523, May 2012. DOI: 10.1016/j.adhoc.2011.09.006. 14

[61] H. Van Trees, *Detection, Estimation and Modulation Theory*. John Wiley & Sons, Inc., 1968. DOI: 10.1002/0471221082. 14, 15

[62] M. Alsheikh, S. Lin, D. Nyato, and H. Tan, Machine learning in wireless sensor networks: algorithms, strategies, and applications, *IEEE Communications Surveys and Tutorials*, pp. 1553–877x, April 2014. DOI: 10.1109/comst.2014.2320099. 14

[63] P. Jarabo-Amo, R. Rosa-Zurera, R. Gil-Pita, and F. Lopez-Ferreras, Sufficient condition for an adaptive system to approximate neyman-pearson detector, *IEEE 13th Workshop on Statistical Signal Processing*, pp. 295–300, July 2005. DOI: 10.1109/ssp.2005.1628609. 14

[64] S. Kay, *Fundamentals of Statistical Signal Processing: Detection Theory*. Prentice Hall, 1993. 15

[65] J. Proakis, C. Rader, F. Ling, M. Moonen, I. Proudler, and C. L. Nikias, *Algorithms for Statistical Signal Processing*, Pearson, 2002. 17

[66] J. Yi and L. Zhou, Enhanced location algorithm with received-signal-strength using fading kalman filter in wireless sensor networks, *IEEE Conference Publishing*, pp. 458–461, 2011. DOI: 10.1109/iccps.2011.6089930. 18

[67] M. Gholami, H. Wymeersch, E. Strom, and M. Rydstrom, Wireless network positioning as a convex feasibility problem, *EURSIP Journal on Wireless Communications and Networking*, January 2011. DOI: 10.1186/1687-1499-2011-161. 19

[68] A. Hero and D. Blatt, Sensor network source localization via projection onto convex sets, *IEEE International Conference on Acoustics, Speech and Signal Processing*, vol. 3, March 2005. DOI: 10.1109/icassp.2005.1415803. 20

[69] J. Foutz, A. Spanias, and M. K. Banavar, *Narrowband Direction of Arrival Estimation for Antenna Arrays*. Morgan & Claypool Publishers, 2008. DOI: 10.2200/s00118ed1v01y200805ant008. 24

[70] G. Mao, *Localization Algorithms and Strategies for Wireless Sensor Networks*. Information Science Reference, 2009. DOI: 10.4018/978-1-60566-396-8. 24

[71] J. Lee, M. Stanley, A. Spanias and C. Tepedelenlioglu, Machine learning in embedded sensor systems for internet-of-things applications. *Proc. 2016 IEEE International Symposium on Signal Processing and Information Systems (ISSPIT 2016)*, Limassol, Cyprus, Dec. 2016. 2

[72] X. Zhang, M. Banavar, C. Tepedelenlioglu, and A. Spanias, Maximum likelihood localization in the presence of channel uncertainties. U.S. Patent No. 9,507,011, Patent Issued Nov. 2016. 12

[73] S. Zhang, J. Lee, C. Tepedelenlioglu, and A. Spanias, Distributed estimation of the degree distribution in wireless sensor networks. *IEEE Global Communications Conference*, Dec. 2016. 12

[74] S. Zhang, C. Tepedelenlioglu, M. K. Banavar and A. Spanias, Distributed node counting in wireless sensor networks. *49th Annual Asilomar Conference on Signals, Systems, and Computers*, 2015. 33

[75] S. Zhang, C. Tepedelenlioglu, M. K. Banavar, and A. Spanias, Max-consensus using soft maximum, *47th Annual Asilomar Conference on Signals, Systems, and Computers*, 2013. 33

[76] S. Zhang, C. Tepedelenlioglu, M. K. Banavar and A. Spanias, Max consensus in sensor networks: Non-linear bounded transmission and additive noise, *IEEE Sensors Journal*, vol. 16, pp. 9089–9098, Dec. 2016. 5, 12, 33

[77] A. Savvides, L. Girod, M. B. Srivastava, and D. Estrin, Localization in sensor networks, in *Wireless Sensor Networks*, C. S. Raghavendra, K. M. Sivalingam, and T. Znati, Eds., Norwell, MA, Kluwer, 2004. 34

[78] N. Bulusu, J. Heidemann, and D. Estrin, GPS-less low cost outdoor localization for very small devices, *IEEE Pers. Commun.*, vol. 5, no. 5, pp. 28–34, Oct. 2000.

[79] N. Patwari, A. O. Hero III, M. Perkins, N. S. Correal, and R. J. O'Dea, Relative location estimation in wireless sensor networks, *IEEE Trans. Signal Processing*, vol. 51, no. 8, pp. 2137–2148, Aug. 2003.

[80] G. Han, J. Jiang, C. Zhang, T. Q. Duong, M. Guizani, and G. K. Karagiannidis, A survey on mobile anchor node assisted localization in wireless sensor networks, *IEEE Communications Surveys and Tutorials*, vol. 18, pp. 2220–2243, 2016.

[81] J. Yick, B. Mukherjee, and D. Ghosal, Wireless sensor network survey, *Computer Networks*, Elsevier, 2. 34

[82] J. Foutz+, A. Spanias, S. Bellofiore, and C. Balanis, Adaptive Eigen-projection beamforming algorithms for 1-D and 2-D antenna arrays, *IEEE Antennas and Propagation Letters*, vol. 83, pp. 1929–1935, 2003. 34

[83] S. Bellofiore, J. Foutz+, C. Balanis, A. S. Spanias, T. Duman, and J. Capone, Smart antennas for mobile adhoc networks, *IEEE Trans. on Antennas and Propagation*, vol. 50, no. 5, pp. 571–581, May 2002.

[84] S. Bellofiore, C. Balanis, J. Foutz, and A. S. Spanias, Smart antennas systems for mobile communications networks: Part 2: algorithms, *IEEE Antennas and Propagation Magazine*, vol. 44, no. 4, pp. 106–114, August 2002.

[85] S. Miller and A. Spanias, Algorithms for quotient control in beamforming, *IEEE Antennas and Wireless Propagation Letters*, vol. 6, pp. 651–654, December 2007.

[86] S. Haykin, J. Litva, and T. J. Shepherd, *Radar Array Processing*, Springer-Verlag, New York, 1993.

[87] M. Miller and D. Fuhrmann, Maximum likelihood narrow-band direction finding and the EM algorithm. *IEEE Trans. Acoust. Speech*, 38, pp. 1560–1577, 1990.

[88] P. Stoica and A. Gershman, Maximum-likelihood DOA estimation by data-supported grid search. *IEEE Signal Proc. Let.*, 6, pp. 273–275, 1999.

[89] B. Ottersten, M. Viberg, P. Stoica, and A. Nehorai, Exact and large sample maximum likelihood techniques. *Radar Array Processing*, Springer-Verlag, New York, 1993.

[90] M. Viberg and A. L. Swindlehurst, A Bayesian approach to auto-calibration for parametric array signal processing, *IEEE Trans.*

[91] A. Manikas, A new general global array calibration method, *ICASSP*, vol. 4. pp. 73–76, 1994.

[92] A. Manikas, *Differential Geometry in Array Processing*, Imperial College Press, 2004. 24, 34

[93] F. Zhao, W. Yao, C. Logothetis, Y. Song, Comparison of super-resolution algorithms for TOA estimation in indoor IEEE 802.11 wireless LANs, *Wireless Communications, Networking and Mobile Computing*, WiCOM 2006. 34

[94] K. Pahlavan, Li Xinrong, and J. P. Makela, Indoor geolocation science and technology, *IEEE Communications Magazine*, vol. 40, Issue 2, pp. 112–118, Feb. 2002.

[95] Z. Luo and P. S. Min, Survey of target localization methods in wireless sensor networks, *19th IEEE International Conference on Networks (ICON)*, 2013.

[96] S. Hara, D. Anzai, T. Yabu, L. Kyesan, T. Derham, and R. Zemek, A perturbation analysis on the performance of TOA and TDOA localization in mixed LOS/NLOS environments, *IEEE Transactions on Communications*, vol. 61, no. 2, pp. 679–689, Feb. 2013.

[97] Trung-Kien Le and Nobutaka Ono, Reference-distance estimation approach for TDOA-based source and sensor localization. *Proc. IEEE ICASSP 2015*, pp. 2549–2553, 2015.

[98] Dan Ohev Zion and Hagit Messer, Envelope only TDOA estimation for sensor network self calibration, *IEEE 8th Sensor Array and Multichannel Signal Processing Workshop (SAM)*, pp. 229–232, 2014. 34

[99] K. Langendoen, N. Reijers, Distributed localization in wireless sensor networks: A quantitative comparison, *Comput. Netw.*, vol. 43. 34

[100] M. Ben Jamaa, K. Anis, and K, Yasir, Easyloc: Rss-based localization made easy, *Procedia Computer Science*, vol. 10, pp. 1127–1133, 2012. 34

[101] M. Nilsson, J. Rantakokko, M. A. Skoglund, and G. Hendeby, Indoor positioning using multi-frequency RSS with foot-mounted INS, *International Conference on Indoor Positioning and Indoor Navigation (IPIN)*, pp. 177–186, 2014.

[102] S. Uluskan and T. Filik, A survey on the fundamentals of RSS based localization, *24th Signal Processing and Communication Application Conference (SIU)*, pp. 1633–1636, 2016. 34

[103] A. Engel and A. Koch, Heterogeneous wireless sensor nodes that target the internet of things. *IEEE Micro*, vol. 36, Issue 6, pp. 8–15, 2016. 34

[104] I. Ahriz and D. Le Ruyet, Greedy probabilistic approach for localization in IoT context, *10th International Conference on Information, Communications and Signal Processing (ICICS)*, 2015.

[105] K. Akkaya, I. Guvenc, R. Aygun, N. Pala, and A. Kadri, IoT-based occupancy monitoring techniques for energy-efficient smart buildings, *IEEE Wireless Communications and Networking Conference Workshops (WCNCW)*, pp. 58–63, 2015. 34

[106] F. Hartmann, K. Worms, F. Pistorius, M. Wanjek, and W. Stork, Energy aware, two-staged localization concept for dynamic indoor environments, *Smart SysTech. European Conference on Smart Objects, Systems and Technologies*, 2016. 34

[107] J. D. Poston, J. Schloemann, R. M. Buehrer, V. V. N. Sriram Malladi, A. G. Woolard, and P. A. Tarazaga, Towards indoor localization of pedestrians via smart building vibration sensing, *International Conference on Location and GNSS (ICL-GNSS)*, 2015. 34

[108] D. Hollosi, G. Nagy, R. Rodigast, S. Goetze, and P. Cousin, Enhancing wireless sensor networks with acoustic sensing technology: Use cases, applications and experiments, *2013 IEEE International Conference on Green Computing and Communications and IEEE Internet of Things and IEEE Cyber, Physical and Social Computing*, 2016. 34

[109] S. Benton, A. Spanias, K, Tu, H. Thornburg, G. Qian, and T. Rikakis, Proceedings of the 3rd IASTED, *International Conference on Signal Processing, Pattern Recognition, and Applications*, vol. 2006, pp. 147–151, Feb. 2016. 34

[110] W. Nuo, S. Ming-Lei, Y. Ming, Y. Yuanyuan, and X. Jiyo, A localization algorithm for underwater wireless sensor networks with the surface deployed mobile anchor node, *6th International Conference on Intelligent Systems Design and Engineering Applications (ISDEA)*, 2015. 34

[111] P.-V. Mekikis, G. Athanasiou, and C. Fischione, A wireless sensor network testbed for event detection in smart homes, *IEEE International Conference on Distributed Computing in Sensor Systems*, 2013. 34

[112] L. Cheng, C. Wu, Y. Zhang, and Li Chen, A rescue-assist wireless sensor networks for large building, *IEEE 8th Conference on Industrial Electronics and Applications (ICIEA)*, pp. 1424–1428, 2013. 34

[113] A. Schwarzmeier, J. Besser, R. Weigel, G. Fischer, and D. Kissinger, Compact back-plaster sensor node for dementia and Alzheimer patient care, *IEEE Sensors Applications Symposium (SAS)*, pp. 75–78, 2014. 34

[114] M. F. M. Colunas, J. M. A. Fernandes, I. C. Oliveira, and J. P. S. Cunha, DroidJacket: An android-based application for first responders monitoring, *6th Iberian Conference on Information Systems and Technologies (CISTI)*, 2011. 34

[115] K. Lorincz, D. J. Malan, T. R. F. Fulford-Jones, A. Nawoj, A. Clavel, V. Shnayder, G. Mainland, M. Welsh, and S. Moulton, Sensor networks for emergency response: Challenges and opportunities, *IEEE Pervasive Computing*, vol. 3, Issue 4, pp. 16–23, 2014.

[116] V. Kumar, D. Rus, and S. Singh, Robot and sensor networks for first responders, *IEEE Pervasive Computing*, vol. 3, Issue 4, 2013. 34

[117] S. Haykin, K. J. R. Liu, *Handbook on Array Processing and Sensor Networks*, vol. 63, John Wiley & Sons, 2010. 34

[118] D. Culler, D. Estrin, and M. Srivastava, Overview of sensor networks, *Computer*, vol. 32, pp. 41–50, 2004.

[119] D. Estrin, L. Girod, G. Pottie, and M. Srivastava, Instrumenting the world with wireless sensor networks, *IEEE ICASSP*, vol. 4, pp. 2033–2036, 2001.

[120] M. Banavar, J. Zhang, B. Chakraborty, H. Kwon, Y. Li, H. Jiang, A. Spanias, C. Tepedelenlioglu, C. Chakrabarti, and A. Papandreou-Suppappola, An overview of recent advances on distributed and agile sensing, algorithms and implementation, *Digital Signal Processing*, Elsevier, 2015. 34

[121] J.-S. Bilodeau, D. Fortin-Simard, S. Gaboury, B. Boucha, and A. Bouzouane, A practical comparison between filtering algorithms for enhanced RFID localization in smart environments, *6th International Conference on Information, Intelligence, Systems and Applications (IISA)*, 2015. 34

[122] T. Sanpechuda and L. Kovavisaruch, A review of RFID localization: Applications and techniques, *Proc. of the 5th International Conference on Electronics Computer Telecommunications and Information Technology IEEE*, pp. 769–772, 2008. 34

[123] C. Tepedelenlioglu, M. K. Banavar, and A. Spanias, On the asymptotic efficiency of distributed estimation systems with constant modulus signals over multiple-access channels, *IEEE Transactions on Information Theory*, vol. 57, no. 10, pp. 7125–7130, Oct. 2011. 34

[124] M. K. Banavar, C. Tepedelenlioglu, and A. Spanias, Estimation over fading channels with limited feedback using distributed sensing, *IEEE Transactions on Signal Processing*, vol. 58, no. 1, pp. 414–425, January 2010.

[125] A. Ribeiro and G. Giannakis, Bandwidth-constrained distributed estimation for wireless sensor networks-part I: Gaussian case, *IEEE Transactions on Signal Processing*, vol. 54, no. 3, pp. 1131–1143, Mar. 2006. 34

[126] M. Banavar, C. Tepedelenlioglu, and A. Spanias, Robust consensus in the presence of impulsive channel noise, *IEEE Trans. on Signal Processing*, vol. 63, pp. 2118–2129, March 2015. 33

[127] G. Mateos, I. D. Schizas, and G. B. Giannakis, Distributed recursive least-squares for consensus-based in-network adaptive estimation, *IEEE Transactions on Signal Processing*, vol. 57, no. 11, pp. 4583–4588, 2009. 33

[128] S. Zhang, C. Tepedelenlioglu, M. K. Banavar, and A. Spanias, Distributed node counting in wireless sensor networks in the presence of communication noise, *IEEE Sensors Journal*, 2017. 33

[129] X. Ou, X. Wu, X. He, Z. Chen, and Qun-ai Yu, An improved node localization based on adaptive iterated unscented Kalman filter for WSN, *IEEE 10th Conference on Industrial Electronics and Applications (ICIEA)*, pp. 393–398, 2015. 34

[130] N. Kovvali, M. Banavar, and A. Spanias, An introduction to Kalman filtering with MATLAB examples, *Synthesis Lectures on Signal Processing*, J. Mura, Ed., vol. 6, no. 2, pp. 1–81, Morgan & Claypool Publishers, ISBN 13: 9781627051392, September 2013.

[131] X. Wang, H. Zhang, and M. Fu, Collaborative target tracking in WSNs using the combination of maximum likelihood estimation and Kalman filtering, *Journal of Control Theory and Applications*, vol. 11, no. 1, pp. 27–34, 2013. 34

[132] H. Kwon, V. Berisha, A. Spanias, and V. Atti+, Experiments with sensor motes and java-DSP, *IEEE Tran. on Education*, vol. 52, issue 2, pp. 257–262, 2009. 34

[133] W. E. Verreycken, D. Laurijssen, W. Daems, and J. Steckel, Firefly based distributed synchronization in wireless sensor networks for passive acoustic localization, *International Conference on Indoor Positioning and Indoor Navigation (IPIN)*, 2016.

[134] M. Amarlingam, P. Rajalakshmi, M. Yoshida, and K. Yoshihara, Mobile phone based acoustic localization for wireless sensor networks, *IEEE 2nd World Forum on Internet of Things (WF-IoT)*, pp. 658–662, 2015.

[135] B. Robistow, R. Newman, T. DePue, M. Banavar, D. Barry, P. Curtis, and A. Spanias, Reflections, an emodule for echo location education, *IEEE ICASSP*, New Orleans, 2017.

[136] P. Misra, S. S. Kanhere, S. Jha, and W. Hu, Sparse representation-based acoustic rangefinders: From sensor platforms to mobile devices, *Communications Magazine IEEE*, vol. 53, no. 1, pp. 249–257, January 2015.

[137] K. D. Frampton, Acoustic self-localization in a distributed sensor network, *Sensors Journal IEEE*, vol. 6, no. 1, pp. 166–172, Feb. 2006. 34

[138] E. D. Nerurkar and S. I. Roumeliotis, A communication-bandwidth-aware hybrid estimation framework for multi-robot cooperative localization, *IEEE/RSJ International Conference on Intelligent Robots and Systems*, pp. 1418–1425, 2013. 34

[139] S. Petridou, S. Basagiannis, and M. Roumeliotis, Survivability analysis using probabilistic model checking: A study on wireless sensor networks, *IEEE Systems Journal*, vol. 7, pp. 4–12, 2013. 34

[140] A. Tahat, G. Kaddoum, S. Yousefi, S. Valaee, and F. Gagnon, A look at the recent wireless positioning techniques with a focus on algorithms for moving receivers, *IEEE Access*, vol. 4, pp. 6652–6680, 2016.

[141] R. Di Taranto, S. Muppirisetty, R. Raulefs, D. Slock, T. Svensson, and H. Wymeersch, Location-aware communications for 5G networks: How location information can improve scalability latency and robustness of 5G, *IEEE Signal Process. Mag.*, vol. 31, no. 6, pp. 102–112, Nov. 2014. 2

[142] S. Yousefi, X.-W. Chang, and B. Champagne, Mobile localization in non-line-of-sight using constrained square-root unscented Kalman filter, *IEEE Trans. Veh. Technol.*, vol. 64, no. 5, pp. 2071–2083, May 2015.

[143] L. Cong and W. Zhuang, Hybrid TDOA/AOA mobile user location for wideband CDMA cellular systems, *IEEE Trans. Wireless Commun.*, vol. 1, no. 3, pp. 439–447, Jul. 2002.

[144] P. Mirowski, T. Kam Ho, S. Yi, and M. MacDonald, SignalSLAM: Simultaneous localization and mapping with mixed WiFi, Bluetooth, LTE and magnetic signals, *International Conference on Indoor Positioning and Indoor Navigation*, pp. 1–10, 2013.

[145] M. Mazzola, G. Schaaf, A. Stamm, and T. Kuerner, Safety-critical driver assistance over LTE: Towards centralized ACC, *IEEE Transactions on Vehicular Technology*, Issue 99, October 2016. 2

[146] M. Winkler, K.-D. Tuchs, K. Hughes, and G. Barclay, Theoretical and practical aspects of military wireless sensor networks, *J. Telecommun. Inf. Technol.*, no. 2, pp. 37–45, Jun. 2008. 34

[147] J. R. Lowell, Military applications of localization tracking and targeting, *IEEE Wireless Commun.*, vol. 18, no. 2, pp. 60–65, Apr. 2011. 34

[148] A. Spanias, *Digital Signal Processing; An Interactive Approach*, 2nd ed., Ch. 9, ISBN 978-1-4675-9892-7, Lulu Press on-demand Publishers Morrisville, NC, May 2014. 34

[149] F. Seco, A. R. Jimenez, C. Prieto, J. Roa, and K. Koutsou, A survey of mathematical methods for indoor localization, *Proc. IEEE Int. Symp. Intell. Signal Process.*, pp. 9–14, Aug. 2009. 34

[150] Y. Gu, A. Lo, and I. Niemegeers, A survey of indoor positioning systems for wireless personal networks, *IEEE Commun. Surveys Tut.*, vol. 11, no. 1, pp. 13–32, Mar. 2009.

[151] H. Liu, H. Darabi, P. Banerjee, and J. Liu, Survey of wireless indoor positioning techniques and systems, *IEEE Transactions on Systems, Man, and Cybernetics Part C*, vol. 37, no. 6, pp. 1067–1080, Nov 2007. 34

[152] F. Gustafsson and F. Gunnarsson, Mobile positioning using wireless networks, *IEEE Signal Processing Magazine*, vol. 22, no. 4, pp. 41–53, July 2005. 34

[153] A. H. Sayed, A. Tarighat, and N. Khajehnouri, Network-based wireless location: Challenges faced in developing techniques for accurate wireless location information, *IEEE Signal Process. Mag.*, vol. 22, no. 4, pp. 24–40, Jul. 2005.

[154] S. Gezici, A survey on wireless position estimation, *Wireless Personal Commun.*, vol. 44, pp. 263–282, Feb. 2008.

[155] R. L. Moses, D. Krishnamurthy, and R. Patterson, A self-localization method for wireless sensor networks, *EURASIP J. Applied Sig. Proc.*, no. 4, pp. 348–358, Mar. 2003.

[156] A. O. Hero and C. M. Kreucher, Network sensor management for tracking and localization, *10th International Conference on Information Fusion*, 2007. 34

Authors' Biographies

XUE ZHANG

Xue Zhang is a Senior System Engineer at Intel Corporation in Santa Clara, CA. She received her B.S. degree from Xi'an Shiyou University in 2008, her M.S. degree from California State University Fullerton in 2010, and her Ph.D. degree from Arizona State University in 2016, all in Electrical Engineering. Her research interests include Digital Signal Processing and Communications. Specifically, she is interested in localization in wireless sensor networks.

CIHAN TEPEDELENLIOGLU

Cihan Tepedelenlioglu was born in Ankara, Turkey in 1973. He received his B.S. degree with highest honors from Florida Institute of Technology in 1995, and his M.S. degree from the University of Virginia in 1998, both in electrical engineering. From January 1999 to May 2001 he was a research assistant at the University of Minnesota, where he completed his Ph.D. degree in Electrical and Computer Engineering. He is currently an associate professor of electrical engineering at Arizona State University. He was awarded the NSF (early) Career grant in 2001, and has served as an associate editor for several IEEE Transactions including *IEEE Transactions on Communications*, *IEEE Signal Processing Letters*, *IEEE Transactions on Wireless Communications*, and *IEEE Transactions on Vehicular Technology*. His research interests include statistical signal processing, system identification, wireless communications, estimation and equalization algorithms for wireless systems, multi-antenna communications, OFDM, ultra-wideband systems, distributed detection and estimation, and data mining for PV systems.

MAHESH BANAVAR

Mahesh Banavar is an assistant professor in the Department of Electrical and Computer Engineering at Clarkson University. He received a B.E. degree in telecommunications engineering from Visvesvaraya Technological University, Karnataka, India in 2005, an M.S. degree and a Ph.D. degree, both in electrical engineering, from Arizona State University in 2007 and 2010, respectively. His research area is signal processing and communications, and he is specifically working on wireless communications and sensor networks. He is a member of MENSA and the Eta Kappa Nu honor society.

ANDREAS SPANIAS

Andreas Spanias is a Professor in the School of Electrical, Computer, and Energy Engineering at Arizona State University. He is also the director of the Sensor Signal and Information Processing (SenSIP) center and the founder of the SenSIP industry consortium (now an NSF I/UCRC site). His research interests are in the areas of adaptive signal processing, speech processing, and sensor systems. He and his student team developed the computer simulation software Java-DSP and its award winning iPhone/iPad and Android versions. He is the author of two textbooks: *Audio Processing and Coding* by Wiley and DSP and *An Interactive Approach* (2nd ed.). He served as associate editor of the *IEEE Transactions on Signal Processing* and as General Co-chair of IEEE ICASSP-99. He also served as the IEEE Signal Processing Vice-President for Conferences. Andreas Spanias is co-recipient of the 2002 IEEE Donald G. Fink paper prize award and was elected Fellow of the IEEE in 2003. He served as distinguished lecturer for the IEEE Signal Processing Society in 2004. He is a series editor for the Morgan & Claypool lecture series on algorithms and software.

Printed in the United States
by Baker & Taylor Publisher Services